Animals Around the World

All About Antarctic Seals

Meg Greve

EZ Readers

An Imprint of Mitchell Lane

Creating Young Nonfiction Readers

EZ Readers lets children delve into nonfiction at beginning reading levels. Young readers are introduced to new concepts, facts, ideas, and vocabulary.

Tips for Reading Nonfiction with Beginning Readers

Talk about Nonfiction

Begin by explaining that nonfiction books give us information that is true. The book will be organized around a specific topic or idea, and we may learn new facts through reading.

Look at the Parts

Most nonfiction books have helpful features. Our *EZ Readers* include color photographs and graphic aids, a table of contents, a glossary, and an index. Share the purpose of these features with your reader.

Color Photos and Graphic Aids

A lot of information can be found by "reading" photos, charts, maps, and other graphic aids found within nonfiction texts. Help your reader learn more about the different ways information can be displayed.

Table of Contents

Located at the front of the book, this list shows the big ideas within the text and the page numbers where they can be found.

Glossary

Located at the back of the book, the glossary defines key words and phrases that are related to the topic. These words and phrases can be found in the text in colored type.

Index

Located at the back of the book, an index is an alphabetical list of topics and the page numbers where they can be found.

With a little help and guidance about reading nonfiction, you can feel good about introducing a young reader to the world of *EZ Readers* nonfiction books.

Mitchell Lane
PUBLISHERS

2001 SW 31st Avenue
Hallandale, FL 33009
mitchelllanepub.com

First Edition, 2026.

Author: Meg Greve
Designer: Rhea Magaro
Editor: Kim Thompson

Names/credits:
Title: All about Antarctic Seals / by Meg Greve
Description: Hallandale, FL :
Mitchell Lane Publishers, [2026]

Series: Animals Around the World
Library bound ISBN: 979-8-89260-520-5
eBook ISBN: 979-8-89260-521-2

Library of Congress Control Number: 2025933683

EZ Readers is an imprint of
Mitchell Lane Publishers

PHOTO CREDITS
Shutterstock: Ondrej Prosicky, cov, 1; Sergey402, 4(top); juan68, 4(bottom); Anton Rodionov, 5(top left); Angela N Perryman, 5(bottom left); Cassandra Sm, 5(top right); Jeremy Richards, 5 (bottom right); chriscasey, 7; Hullis, 8; Szakharov, 10; james_stone76, 12; Phillip Salveter, 14; SichKarenko.com, 16; Ivan Hoerman, 18; Wirestock Creators, 21; MariUSZ, 22(bottom); francesco de marco, 22(top)deer boy, 22(map);

Contents

Antarctica and the Southern Ocean that surrounds it are very cold and windy. Not many animals live here. But seals do! Six kinds of seals make their homes on and around this freezing continent.

Ross Seal

Weddell Seal

Crabeater Seal

Leopard Seal

Fur Seal

Southern Elephant Seal

Seals are **mammals**. Fur covers their bodies. It can be tan, brown, gray, white, or black. Four flippers help seals swim in water and climb on land. A thick layer of **blubber** under their skin helps seals stay warm.

Seals have big eyes to help them see in deep, dark water. They have **sensitive** whiskers. Their whiskers sense anything that moves underwater. This lets seals know right away if a tasty fish or a **predator** is swimming nearby.

Seals are deep divers. Their sleek shape helps them move swiftly through the water. They use their sharp teeth to catch and eat fish and squid. The leopard seal also hunts penguins and seabirds.

Seals spend most of their lives in the water, but they cannot breathe underwater. They must stick their noses above the surface for air. Seals even sleep in the water. They are able to come up for air without waking up!

Seals live alone most of the time. They swim many miles to find food. But at a few special times each year, seals climb up on land or ice. They come together in big, noisy groups called **colonies**.

Seals form colonies on land when it is time to **molt**. They shed all their fur. New, healthy fur grows in its place.

Seals also form colonies when it is time to have babies. Each mother gives birth to one baby. Babies drink their mothers' rich, fatty milk. They grow very fast.

Mothers teach their babies how to swim. They go out into the water and call to the baby. When the baby feels brave enough, it wriggles into the freezing water.

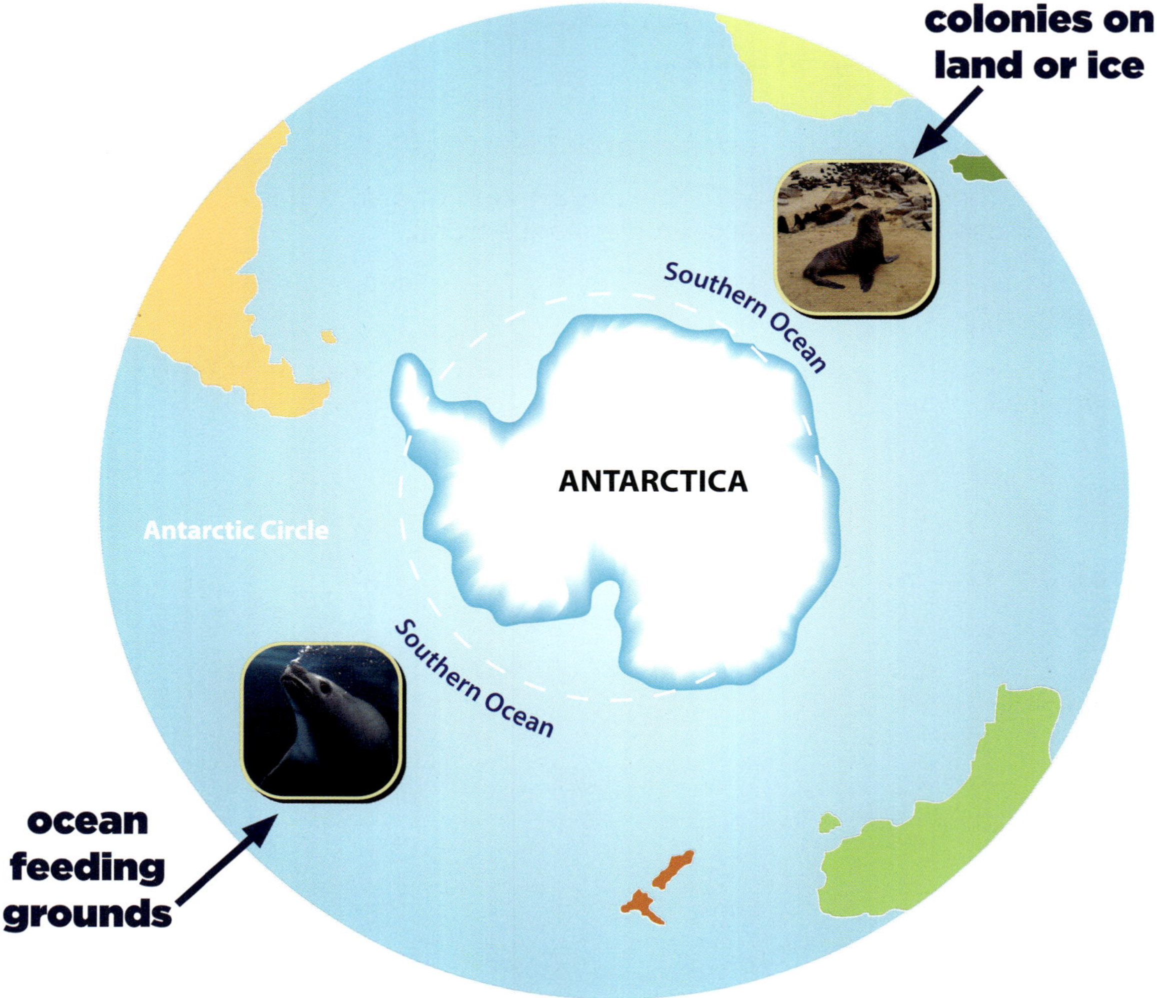

Where Do Seals Live?

Seals live all around the world. Six types of seals swim, dive, and hunt in the Southern Ocean that surrounds the continent of Antarctica. They form colonies on land or ice.

Interesting Facts

- Seals are *pinnipeds*. The word means "fin-footed."
- The Southern elephant seal is the largest kind of seal. It has a big round nose that can look like an elephant's trunk.
- Crabeater seals do not eat crab. They eat tiny ocean animals called krill.
- Seals bark, sing, grunt, groan, click, and splash to communicate with each other.

Parts of a Seal

claws
Claws on their flippers help seals scratch, hold themselves up, and steer in the water.

ears
Seals hear well, but their ears are not easily seen. Seals can close their ears underwater.

flippers
Flippers help seals swim in water and move around clumsily on land.

fur
Short, thick fur helps seals stay warm in icy water.

nostrils
A seal can open and close its nostrils.

Glossary

blubber (BLUHB-ur)
A layer of fat under the skin of an animal that lives in a cold place

colonies (KAH-luh-nees)
Large groups of animals

mammals (MAM-uhlz)
Warm-blooded animals that have backbones and fur or hair, that give birth to live babies, and that feed their babies milk from the mother

molt (mohlt)
To lose old fur, feathers, shells, or skin so that new body parts can grow

predator (PRED-uh-tur)
An animal that hunts other animals for food; a seal's predators include killer whales and sharks

sensitive (SEN-si-tiv)
Able to feel or notice something quickly and easily

Index

Further Reading

Catala, Ellen. *Seals (Active Minds: Kids Ask About)*. Sequoia Kids Media, 2023.

Bowman, Lucy. *Antarctica (Usborne Beginners).* Usborne, 2025.

On the Internet

PBS Nature: Seal Fact Sheet
www.pbs.org/wnet/nature/blog/seal-fact-sheet
Read facts about seals and see beautiful photos of these amazing animals.

World Animal Protection: Five Fun Facts about Seals
www.worldanimalprotection.us/latest/blogs/5-fun-facts-about-seals
Learn five fascinating facts about seals.